2020
The Plandemic Year
Paco Baca.

Foreword.

On October 18, 2019, the Johns Hopkins Center for Health Security, in partnership with the World Economic Forum and the Bill & Melinda Gates Foundation, held "Event 201" in New York.

This event, attended by 130 people, and in which world business, government and public health leaders participated, consisted of a high-level pandemic drill in which areas were exposed to deal with a possible coronavirus pandemic and reduce the large-scale economic and social consequences. "The exercise served to highlight the preparedness and response challenges that would likely arise in a very severe pandemic."

Event 201 simulates an outbreak of a novel zoonotic coronavirus transmitted from bats to pigs to people that eventually becomes efficiently transmissible from person to person, leading to a severe pandemic. The pathogen and the disease it cause are largely based on SARS, but it is more transmissible in the community by people with mild symptoms.

The name "event 201" refers to a severe pandemic that would require cooperation between various industries, national governments and key international institutions.

In their presentation, they explained that the studies carried out at that time showed that pandemics would be the cause of an annual economic loss of 0.7% of world GDP, some 570 billion dollars, on average. In order to act against this, the experts recommended the need to establish ways of cooperation between the industry, national governments, key international institutions and civil society. They then announced that "the next major pandemic will not only cause major illness and loss of life, but could also unleash major cascading economic and social consequences that could greatly contribute to global impact and suffering."

The "Event 201" pandemic exercise demonstrated in real life some of these important gaps in pandemic preparedness, as well as some of the elements of public-private sector solutions that will be needed to fill them." That is, to structure a new economic paradigm that involves health spending, focused on a planetary emergency scheme, which must be self-financed by the countries of the world, to the medical laboratories selected to carry out the research and development tasks. of cures and alternatives to new biological threats and the consequent global pandemics.

Since the entire human population is susceptible, during the first months of the pandemic, the cumulative number of cases increases exponentially, doubling every week. And as cases and deaths mount, the economic and social consequences become increasingly dire.
The proposed scenario ends after 18 months with 65 million deaths. At that time the

pandemic begins to abate due to the decreasing number of susceptible people. The pandemic will continue until there is an effective vaccine or until 80-90% of the world's population has been exposed. From then on, it is likely to be an endemic childhood disease. Already at that time, October 19, it was pointed out that "during a severe pandemic it is likely that the efforts of the public sector to control the outbreak will be overwhelmed." But industry assets, they noted, if implemented quickly and properly, could help save lives and reduce economic losses.

They cited as an example the need for companies with operations focused on logistics, social media, or delivery systems to enable government emergency response, risk communication, and medical countermeasure distribution efforts during the pandemic. As we can see, all these recommendations existed, they were on the table just a few months ago.

And also, in this conference participated the World Economic Forum, among other organizations. Therefore, a roadmap was already available for what would begin a couple of months later.

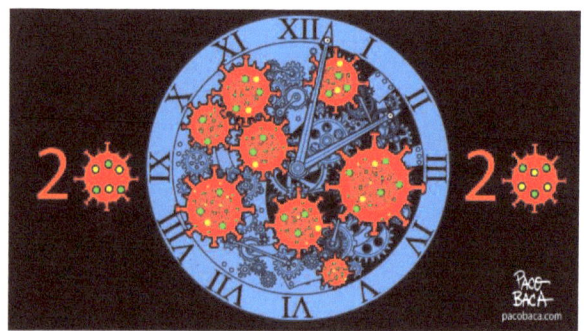

The first case of covid19.

December 31, 2019

The WHO Office in the People's Republic of China notes a statement by the Wuhan Municipal Health Commission to the media published on its website mentioning cases of "viral pneumonia" in Wuhan (People's Republic of China).

January 1, 2020

The WHO asks the Chinese authorities for information about the cluster of cases of atypical pneumonia in Wuhan of which it has received news.

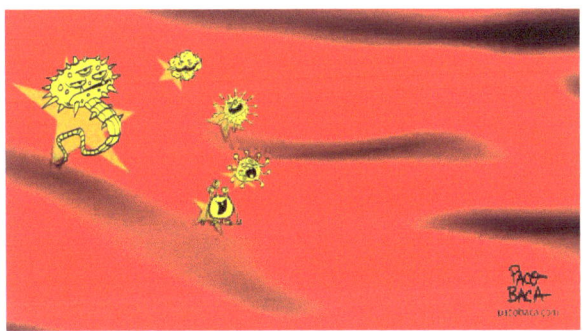

January 2, 2020

The WHO Representative in China writes to the National Health Commission to offer the Organization's support and repeat the request for more information on the cluster of cases.

January 3, 2020

The WHO receives information from Chinese officials about the cluster of cases of "viral pneumonia of unknown origin" detected in Wuhan.

January 4, 2020

The WHO publishes on Twitter that there was a conglomerate of pneumonia cases – without fatalities – in Wuhan, Hubei province (People's Republic of China) and that investigations had begun to determine the cause

January 5, 2020

WHO shares detailed information on a cluster of cases of pneumonia of unknown cause through the IHR Event Information System (2005), which is accessible to all Member States.

The event notice provided information on cases and advised Member States to take precautions to reduce the risk of acute respiratory infections.

January 9th

The WHO reports that the Chinese authorities have determined that the outbreak is caused by a new coronavirus.

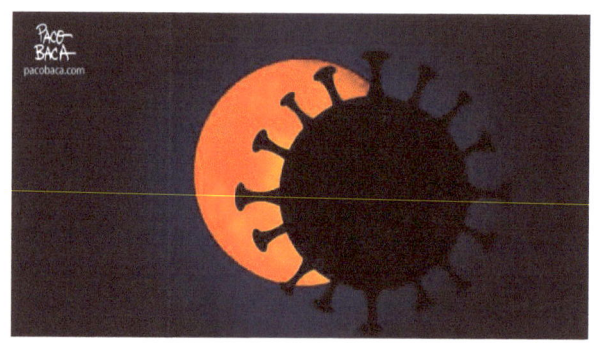

January 10, 2020

The Global Coordination Mechanism for research and development activities for epidemic prevention and response holds its first teleconference on the novel coronavirus.

January 10-12, 2020

WHO publishes a comprehensive set of guidance documents for countries on issues related to managing a new disease outbreak:

January 11, 2020

The WHO publishes on Twitter that it had received the genetic sequences of the new coronavirus from the People's Republic of China and that it hoped they would be published soon.

Chinese media reports the first fatality from the new coronavirus.

January 12, 2020

WHO convenes the first teleconference with the global network of experts in diagnostics and laboratories.

January 13, 2020

Thailand's Minister of Public Health reports a case of the laboratory-confirmed novel coronavirus imported from Wuhan, the first recorded case outside the People's Republic of China.

The WHO publishes the first protocol of the retro-transcription polymerase chain reaction (RT-PCR) test by a partner laboratory of the Organization to diagnose the new coronavirus.

January 14, 2020

The WHO convenes a press conference in which it states that, based on experience with respiratory pathogens, there is a risk of possible human-to-human transmission in the 41 confirmed cases in the People's Republic of China: "it is certainly possible that producing limited human-to-human transmission.

January 16, 2020

Japan's Ministry of Health, Labor and Welfare notifies WHO of a confirmed case of novel coronavirus infection in a person who had traveled to Wuhan. It is the second confirmed case detected outside the People's Republic of China. WHO states that given international travel patterns, more cases were likely,

January 17, 2020

WHO convenes the first meeting of the analysis and modeling working group for the new coronavirus

January 19, 2020

The WHO Western Pacific Regional Office (WPRO) posts on Twitter that, based on the latest information received and WHO analysis, there was evidence of limited human-to-human transmission

January 20, 2020

WHO publishes guidance on home care for suspected infected patients.

January 20-21, 2020

WHO carries out the first mission to Wuhan and meets with public health officials to gather information on the response to the cluster of cases of novel coronavirus infection.

January 21, 2020

WPRO posts on Twitter that there is now no doubt from the most recent information that "at least some human-to-human transmission" occurs, and that infections among health professionals corroborate this. The United States of America (USA) reports its first confirmed case of infection with the novel coronavirus. This is the first case in the WHO Region of the Americas. WHO convenes the first meeting of the global network of experts on infection prevention and control.

January 22, 2020

The WHO mission to Wuhan issues a statement stating that scientific data points to inevitable human-to-human transmission in Wuhan.

January 22-23, 2020

The Director General convenes an Emergency Committee under the International Health Regulations (IHR) on the outbreak of the novel coronavirus. The Committee, made up of 15 independent experts from different parts of the world, was mandated to advise the Director General whether the outbreak constituted a Public Health Emergency of International Concern (ESPII).

January 24, 2020

France notifies the WHO of three cases of infection with the new coronavirus, all from people who had traveled from Wuhan. These are the first confirmed cases in the WHO European Region (EURO).

The Director of the Pan American Health Organization (PAHO) urges the countries of the Americas to be prepared to early detect, isolate, and care for patients infected with the new coronavirus, given the possibility of receiving travelers from countries where there is transmission of the virus. new coronavirus.

January 25, 2020

WHO Regional Director for Europe issues a public statement highlighting the importance of preparing at local and national levels to detect cases, analyze samples and provide clinical care.

January 27, 2020

The WHO Regional Director for South-East Asia issues a press release urging countries in the Region to focus on being prepared to rapidly detect imported cases and prevent further spread of the virus.

January 27-28, 2020

A delegation of senior WHO officials led by the director-general arrives in Beijing to meet Chinese leaders, learn more about the response in the PRC, and offer technical assistance.

The CEO meets with Xi Jinping, President of China, on January 28, with whom he discusses continued collaboration on containment measures in Wuhan.

January 29, 2020
On his return to Switzerland from China, the Director-General provides Member States with an update on the response to the outbreak of the novel coronavirus infection in China, during the 30th meeting of the Program,

Budget and Administration Committee of the Executive Board.

Case numbers outside of China point to the possibility of a much larger outbreak, although they were still relatively low at the time.

The United Arab Emirates reports the first cases in the Eastern Mediterranean Region.

January 30, 2020

WHO is holding a briefing with Member States to provide them with more information on the outbreak.

The WHO publishes a guide to measures on the use of masks in the community setting, in home care and in health centers.

At that time there were 98 cases and no fatalities in 18 countries outside of China. Four countries had evidence (8 cases) of person-to-person transmission outside of China (Germany, Japan, the United States of America, and Viet Nam).

January 31, 2020

The WHO Regional Director for Africa sends all countries in the Region a guidance notes emphasizing the importance of preparedness and early detection of cases.

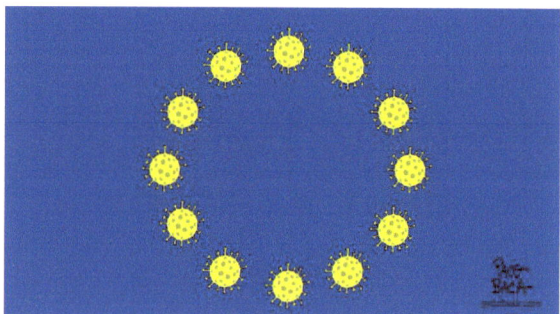

February 2, 2020

First batch of diagnostic kits for laboratory RT-PCR testing sent to WHO regional offices.

February 3, 2020

WHO finalizes its Strategic Preparedness and Response Plan, focused on improving the ability to detect, prepare for and respond to the outbreak.

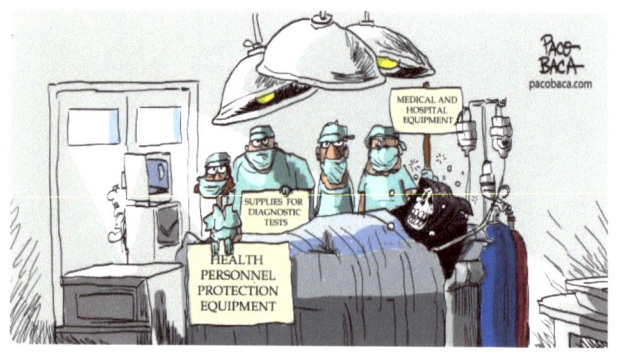

February 4, 2020

The WHO director-general calls on the UN secretary-general to activate the UN crisis management team, which holds its first meeting on February 11.

In response to a question posed in the Executive Council, the Secretariat replies that "it is possible that asymptomatic people spread the virus, but we need more detailed studies on the matter to determine the frequency and if it leads to secondary transmission."

February 5, 2020

At the WHO headquarters, daily press conferences on the new coronavirus begin to be held. This is the first time that the Director General or Executive Director of the WHO Health Emergencies Program has offered daily briefings.

February 9, 2020

WHO deploys an advance team of the WHO-China Joint Mission, after receiving final clearance from the People's Republic of China on the same day. The mission had been agreed between the director general and Xi Jinping during the visit of the WHO delegation to China at the end of January.

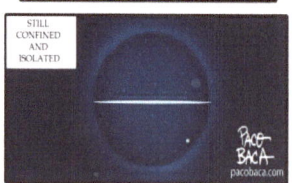

February 11, 2020

The WHO announces that the disease caused by the new coronavirus will be called COVID-19. Observing best practice, that name was chosen to avoid inaccuracies and stigmatization; therefore, it does not refer to a geographical location, an animal, a person or a group of people.

February 11-12, 2020

WHO convenes a Global Research and Innovation Forum on the novel coronavirus, attended by more than 300 experts and funders from 48 countries, plus another 150 online participants.

The Forum is convened in line with the WHO R&D Project, which was activated to accelerate diagnostics, vaccines and treatments against this new coronavirus.

February 12, 2020

To complement the Strategic Plan with more information, WHO publishes guidelines for operational planning to support country preparedness and response.

February 13, 2020

The WHO Digital Solutions Unit convenes a roundtable of 30 companies in Silicon Valley to help WHO keep people safe and informed about COVID-19.

February 14, 2020

Building on lessons from the H1N1 and Ebola outbreaks, WHO finalizes guidance for organizers of mass gatherings in the context of COVID-19.

February 15, 2020

The CEO delivers a speech at the Munich Security Conference, a global forum dedicated to international security issues, particularly health security, where he also holds several bilateral meetings.

It also expresses its concern about the lack of urgency with which the financing of the response is addressed.

February 16, 2020

The WHO-China Joint Mission begins work. As part of the international mission to assess the severity of this new disease, its transmission dynamics, and the characteristics and effects of China's control measures, teams are conducting field visits to Beijing, Guangdong, Sichuan, and Wuhan.

February 19, 2020

Weekly briefings on COVID-19 for Member States to inform them of the latest data and news on the disease begin.

February 21, 2020

The WHO Director-General appoints six special envoys on COVID-19 to provide strategic advice, high-level political engagement and advocacy in different parts of the world:
• Professor Dr. Maha El Rabbat, former Minister of Health of Egypt;• Dr. David Nabarro,

former Special Adviser to the United Nations Secretary General on the 2030 Agenda for Sustainable Development and Climate Change;
• Dr. John Nkengasong, Director of the African Centers for Disease Control and Prevention;
• Dr. Mirta Roses, former WHO Regional Director for the Americas;
• Dr Shin Young-soo, former WHO Regional Director for the Western Pacific; and
• Professor Samba Sow, Director General of the Vaccine Development Center in Mali.

February 24, 2020

The team leaders of the WHO-China Joint Mission on COVID-19 hold a press conference to share their main findings.

The Mission warns that "a large part of the world community is still not prepared, neither mentally nor materially, to apply the measures that have been implemented in China to contain COVID-19."

February 25, 2020

Confirmation of the first case in the WHO African Region, in Algeria. This case comes after the previous notification of a case in Egypt, the first on the African continent. The regional director for Africa calls on countries to intensify their preparation.

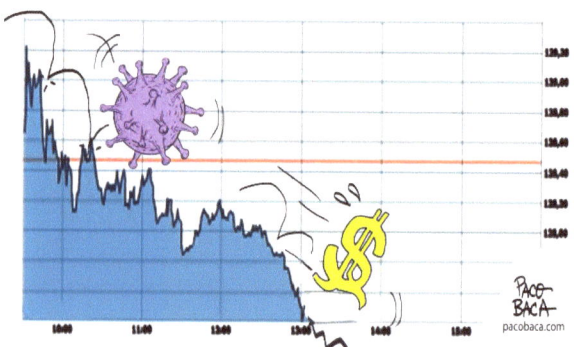

February 27, 2020

WHO publishes guidance on the rational use of personal protective equipment, in view of the global shortage. The guidance provides recommendations on the type of personal protective equipment to be used based on the environment, personnel, and type of activity.

February 28, 2020

The Report of the WHO-China Joint Mission is published, to serve as a reference for countries on the necessary measures to contain COVID-19.

February 29, 2020

The WHO publishes considerations for the quarantine of people in the context of the containment of COVID-19, which indicate the people who should undergo quarantine and the minimum conditions for quarantine to avoid the risk of new transmissions.

March 3, 2020
The WHO is urging industry and governments to increase production by 40% to meet growing global demand in response to shortages of personal protective equipment, shortages that endanger healthcare workers around the world.

March 6, 2020

WHO publishes the Global Research Roadmap against the new coronavirus, prepared by the working groups of the Research Forum.

Focusing on long-term goals for treatments and vaccines.

March 7, 2020

Underlining that the 100,000 confirmed cases of COVID-19 have been surpassed, the WHO issues a statement calling for action to stop, contain, control, delay and reduce the impact of the virus at every opportunity.

March 10, 2020

WHO, UNICEF and the International Federation of Red Cross and Red Crescent Societies publish a guide outlining the most important considerations for keeping schools safe, with practical checklists and advice for parents and careers, as well as for the children and students themselves.

March 11, 2020

Deeply concerned by the alarming levels of spread and severity and by the alarming levels of inaction, the WHO concludes in its assessment that COVID-19 can be considered a pandemic.

March 13, 2020

The Director General declares that Europe has become the epicenter of the pandemic, with more reported cases and deaths than the rest of the world combined, apart from the People's Republic of China.

March 17, 2020

WHO, together with the International Federation of Red Cross and Red Crescent Societies, the International Organization for Migration (IOM) and the United Nations Refugee Agency (UNHCR), publishes guidance on expansion of COVID-19 outbreak preparedness and response operations in camps and similar settings.

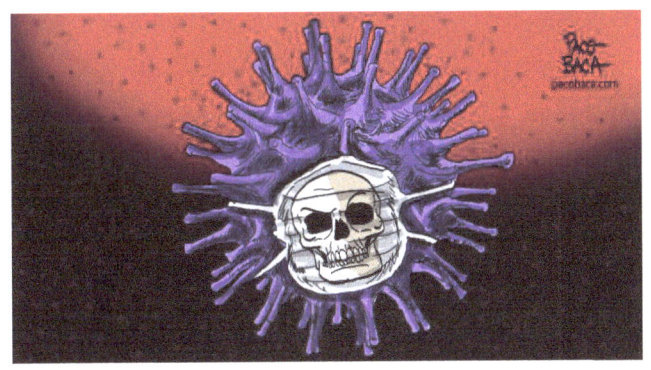

March 18, 2020

WHO and partners launch the Solidarity trial, an international clinical trial that aims to generate rigorous data from around the world to find the most effective treatments against COVID-19.

WHO publishes guidance on mental health and psychosocial considerations during the COVID-19 outbreak.

March 20, 2020

A WHO health alert messaging service is launched on WhatsApp that offers instant and accurate information on COVID-19. It is available in multiple languages and has users all over the world.

March 21, 2020

Given the lack of capacity in many Member States to carry out tests, WHO publishes the recommendations on the strategy for laboratory tests related to COVID-19.

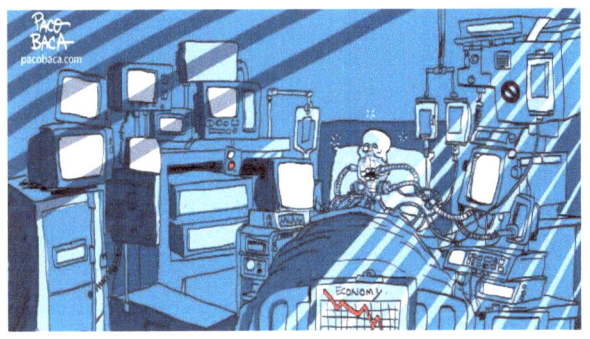

March 23, 2020

WHO and FIFA launch the "Give the message and kill the coronavirus" awareness campaign, spearheaded by well-known international footballers, calling on people around the world to protect their health.

March 25, 2020

The Director-General of WHO, the Secretary-General of the United Nations, the Under-Secretary-General for Humanitarian Affairs of the United Nations and the

Executive Director of UNICEF present the United Nations Global Humanitarian Response Plan.

March 26, 2020

Director-General addresses G20 Extraordinary Summit on COVID-19, chaired by King Salman of Saudi Arabia, urging G20 leaders to fight, unite and innovate against COVID-19

In addition to the G20, WHO joins UNESCO and other partners to launch the Global Education Coalition to provide children and youth with inclusive learning options during this period of sudden and unprecedented disruption to education.

March 28, 2020

With many health facilities around the world overwhelmed by the influx of COVID-19 patients seeking medical care, WHO publishes a manual for setting up and running a severe acute respiratory infection treatment center and screening area of these in medical care establishments in order to optimize patient care.

March 30, 2020

The Director-General urges countries to work with companies to increase production, ensure the free movement of essential health products and ensure the fair distribution of those products, having recently addressed the G20 trade ministers on the ways to solve chronic shortages.

March 31, 2020

WHO issues a Medical Products Alert warning consumer, healthcare professionals and health authorities against the growing number of counterfeit medical products that claim to prevent, detect, treat or cure COVID-19.

April 2, 2020

The WHO presents the evidence of the transmission of symptomatic, pre-symptomatic and asymptomatic people with COVID-19 and points out that transmission can occur from a pre-symptomatic case, that is, before the onset of symptoms.

April 4, 2020

The WHO reports that more than one million cases of COVID-19 have already been confirmed worldwide. In other words, the number of cases has multiplied by ten in less than a month.

April 6, 2020

The WHO updates the recommendations on the use of masks and includes a new section with recommendations for decision-makers on the use of masks by healthy people in the community.

April 7, 2020

World Health Day focuses on commemorating the work of nurses and midwives on the front lines of the response to COVID-19.

April 8, 2020

The United Nations Task Force on Supply Chains is launched to coordinate and scale up the procurement and distribution of personal protective equipment, laboratory diagnostic tests, and oxygen to countries that need it most.

April 9, 2020

WHO marks 100 days since the first cases of "pneumonia of unknown cause" were reported with a synopsis of major developments and efforts to stop the spread of the coronavirus.

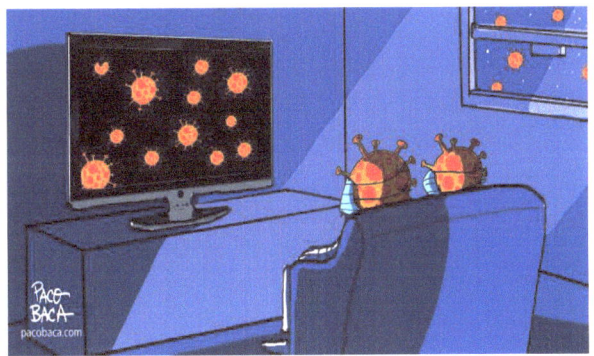

April 11, 2020

WHO publishes a draft overview of vaccine candidates against the virus that causes COVID-19, based on a systematic and constantly updated evaluation of candidates from around the world.

April 13, 2020

WHO publishes a statement signed by 130 scientists, donors and manufacturers from around the world pledging to work with WHO to accelerate the development of a vaccine against COVID-19.

April 14, 2020

WHO publishes COVID-19 Strategy Update, with guidance for countries preparing for a gradual transition from widespread transmission to a steady state of low or no transmission. Its goal is for all countries to control the pandemic by mobilizing all sectors and communities to prevent and suppress community transmission, reduce mortality, and develop safe and effective vaccines and treatments.

April 15, 2020

WHO finalizes guidance with advice on public health and social and religious practices during Ramadan in the context of COVID-19.

April 16, 2020

WHO issues guidance on considerations for adjusting public health and social measures in the context of COVID-19, such as large-scale mobility restrictions, commonly referred to as "lockdown" or "isolation" measures.

April 18, 2020

WHO and Global Citizen are jointly organizing the One World: Together At Home concert, a special live global edition to celebrate and support frontline health workers. The concert raises a total of US$127.9 million, of which US$55.1 million goes to the COVID-19 Solidarity Response Fund and US$72.8 million to those responsible for the response local and regional.

April 19, 2020

Together with 14 other humanitarian organizations, WHO is calling on the donor community to provide urgent support to the global emergency supply system to combat COVID-19.

April 20, 2020

The United Nations General Assembly adopts a resolution entitled "International cooperation to ensure global access to medicines, vaccines and medical equipment to deal with COVID-19".

April 24, 2020

In a virtual event jointly organized by the WHO, Emmanuel Macron, President of France, Ursula Von der Leyen, President of the European Commission, and the Bill & Melinda Gates Foundation, the

Director-General presents the Accelerator for access to tools against COVID -19, a collaboration with the goal of accelerating the development, production, and equitable access to vaccines, diagnostic tests, and treatments against COVID-19.

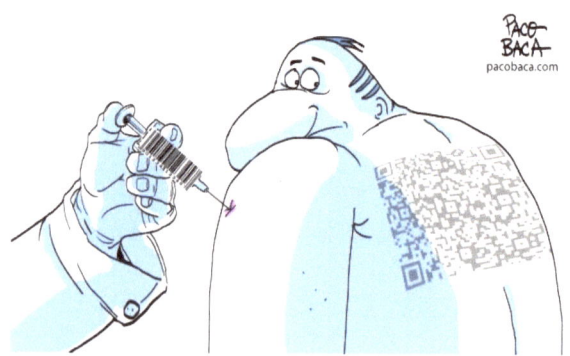

April 30, 2020

The CEO convenes the third meeting of the International Health Regulations Emergency Committee on COVID-19, with a larger membership to reflect the nature of the pandemic and the need to include technical expertise in other areas. The Emergency Committee meets on April 30 and issues its statement on May 1.

May 4, 2020

The Director-General addresses leaders from 40 countries around the world at a pledging event for the global response to COVID-19 hosted by the European Commission.

May 5, 2020

WHO launches the COVID-19 Supply Portal, a purpose-built tool to facilitate and consolidate the submission of supply requests by national authorities and all implementing partners in support of national COVID-19 action plans. COVID-19.

May 7, 2020

The United Nations presents a $6.7 billion Global Humanitarian Response Plan update to minimize the most debilitating effects of the pandemic in 63 low- and middle-income countries.

May 10, 2020

Building on previous guidance on case and cluster investigation, WHO is issuing interim guidance on contact tracing.

May 10-14, 2020

As Member States face different transmission scenarios, WHO publishes four annexes to considerations for adjusting public health and social measures in the workplace, schools and mass gatherings, as well as public health criteria to adjust those measurements.

May 13, 2020

Designed to provide health workers with information to help them care for COVID-19 patients and protect themselves, the WHO Academy mobile app is launched, together with the WHO Info app, for the public in general.

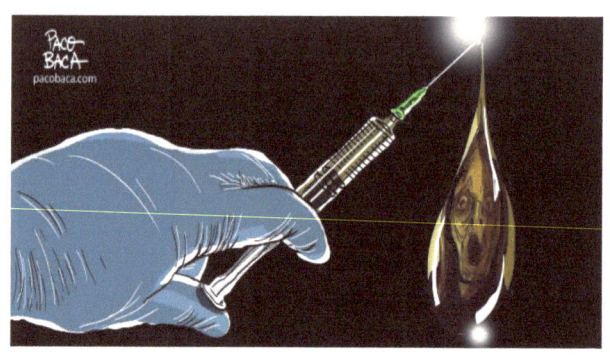

May 14, 2020

WHO publishes a supporting document advising countries to incorporate a gender approach into their responses to COVID-19 in order to ensure that public health policies and measures to curb the epidemic take into account gender and the way it interacts with other inequalities.

May 15, 2020

WHO publishes a scientific report on multisystem inflammatory syndrome in children and adolescents, temporarily related to COVID-19.

May 18, 2020

The Independent Oversight and Advisory Committee for the WHO Health Emergencies Program, which oversees and monitors WHO's work in health emergencies, concludes its interim report on the WHO response to COVID-19 during the period of January to April 2020.

May 18-19, 2020

The 73rd World Health Assembly, the first to be held virtually, passes a landmark resolution to unite the world in the fight against the COVID-19 pandemic, co-sponsored by more than 130 countries—the largest number on record—and adopted by consensus. In the opening and closing sessions, 14 heads of state participate.

May 21, 2020

WHO signs a new agreement with the Office of the United Nations High Commissioner for Refugees (UNHCR). One of the main objectives of the agreement for the year 2020 is to help protect some 70 million forcibly displaced people from COVID-19.

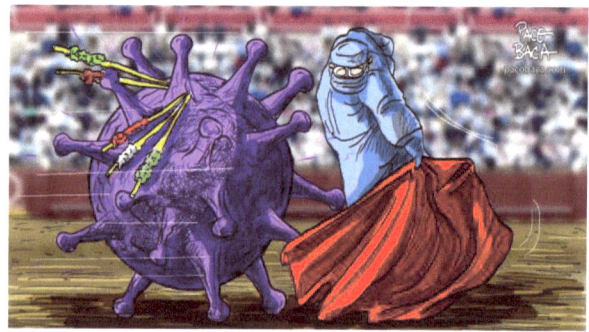

May 27, 2020

The WHO Foundation is established with the goal of supporting global public health needs by contributing funds to WHO and trusted partners.

May 29, 2020

A total of 30 countries and multiple international partners and institutions launch the Pooled Access to Technology against COVID-19 (C-TAP) initiative, which aims to ensure that vaccines, tests, treatments and other health technologies related to the fight against COVID-19 are available to everyone.

June 2, 2020

The Executive Director of the WHO Health Emergencies Program addresses the High-Level Pledging Conference for Yemen, organized to support humanitarian response and alleviate suffering in the country

June 4, 2020

WHO welcomes the crucial new funding commitments made at the World Vaccine Summit. Organized by the UK Government. The Summit also highlights the importance of having a safe and effective vaccine that is equitably accessible to fight COVID-19.

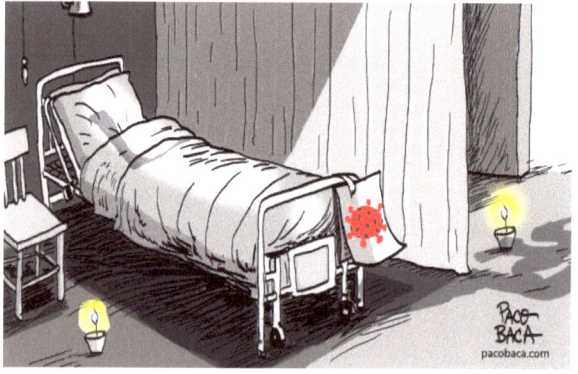

June 5, 2020

WHO updates guidance on the use of masks to control COVID-19, providing up-to-date information on who should wear a mask and when, as well as what materials masks should be made of.

June 13, 2020

The WHO reports that the Chinese authorities have provided information on a cluster of COVID-19 cases in Beijing (People's Republic of China).

WHO offers technical support and assistance and requests more information on the cluster, as well as on ongoing and planned investigations.

June 16, 2020

WHO welcomes initial results from a UK clinical trial indicating that dexamethasone, a corticosteroid, may save the lives of critically ill COVID-19 patients

June 17, 2020

The WHO announces the discontinuation of the hydroxychloroquine treatment group of the Solidarity trial, which aims to find an effective treatment against COVID-19.

June 26, 2020

The Access to COVID-19 Tools Accelerator publishes the consolidated case for investment and calls for an investment of US$ 31.3 billion over the next 12 months in diagnostic tests, treatments and vaccines.

July 27, 2020

WHO marks World Hepatitis Day by highlighting the results of a modeling study conducted in collaboration with Imperial College London.

July 31, 2020

The Committee unanimously agrees that the pandemic continues to constitute a Public Health Emergency of International Concern (ESPII) and offers its advice to the Director General. The CEO declares that the COVID-19 outbreak continues to constitute a PHEIC.

It is also recommended that countries participate in the ACT Accelerator and relevant clinical trials, and prepare for the introduction of safe and effective vaccines and treatments.

August 3, 2020

WHO publishes its report on progress in preparedness and response to COVID-19, covering the achievements made from February 1 to June 30, 2020 in intensifying international coordination and support and preparing the countries.

August 5, 2020

The CEO is hosting the #WearAMask challenge on social media to help spread the word about how and when to wear a mask to protect against COVID-19.

A plane with 20 tons of WHO medical supplies lands in Beirut, Lebanon to support the treatment of patients who have suffered injuries in the huge explosion that occurred in the city on August 4, in the context of the COVID-19 outbreak, recent public disorder, a major economic crisis and a high refugee burden.

August 6, 2020

WHO holds its regular COVID-19 press briefing in conjunction with the Aspen Security Forum, where the Director-General stresses the crucial importance for national security of investing in health, emphasizing that "no country will be safe until we are all safe." The COVAX Facility allows countries to benefit from a pipeline of candidate vaccines so that citizens can have early access to effective vaccines. The Facility is developed through the COVAX pillar of the ACT Accelerator, in which WHO, the Gavi Vaccine Alliance and the Coalition for Epidemic Preparedness Innovations (CEPI) are working together, together with multinationals and vaccine manufacturers in developing countries.

August 7, 2020

WHO publishes updated guidance on public health surveillance for COVID-19, including revised definitions of suspected case and probable case that capture new information on the clinical spectrum of COVID-19 and its transmission.

August 12, 2020

WHO publishes updated guidance on home care for suspected or confirmed cases of COVID-19 and management of their contacts.

August 14, 2020

WHO, the International Narcotics Control Board (INCB) and the United Nations Office on Drugs and Crime (UNODC) issue a statement calling on governments to ensure that the procurement and supply of internationally controlled medicines in countries meet the needs of patients, both those with COVID-19 and those who need them for other conditions.

August 19, 2020

On World Humanitarian Day, WHO joins with United Nations partners to honor frontline workers around the world responding to COVID-19 and other health emergencies.

These heroes without capes include refugees in essential roles as health workers in the response to the pandemic; Ebola health workers mobilizing against COVID-19; and medical and nursing staff who continue to provide crucial medical care for women and children.

August 21, 2020

WHO, in collaboration with UNICEF, publishes guidance on the use of masks by children in the community in the context of COVID-19.

August 27, 2020

At a briefing for Member States, the Director General announces his plan to establish a Review Committee on the operation of the IHR (International Health Regulation) during COVID-19.

August 28, 2020

WHO launches its "Science in 5" video and podcast series, in which WHO experts provide scientific explanations on specific issues related to COVID-19, to help people protect themselves and others the rest. In the first episode, the WHO chief scientist explains the concept of "herd immunity."

August 31, 2020

WHO publishes the results of its first indicative survey on the effects of COVID-19 on health systems, based on information provided by 105 countries. Almost all countries (90%) have experienced interruptions in their health services, with low- and middle-income countries reporting the greatest difficulties.

September 1, 2020

Following the May forum and July webinar with civil society, the first session of a series of COVID-19 "civil society dialogue" meetings with the CEO is held, focusing on achieving a gender-transformative response to COVID-19.

September 2, 2020

WHO publishes guidance on the role of corticosteroids in the treatment of COVID-19, developed in collaboration with the Magic Evidence Ecosystem Foundation (MAGIC), a non-profit foundation.

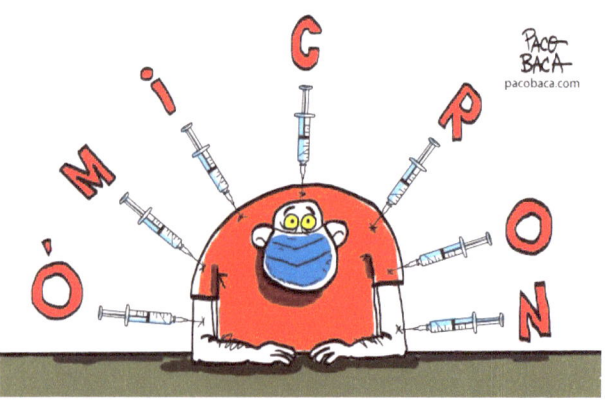

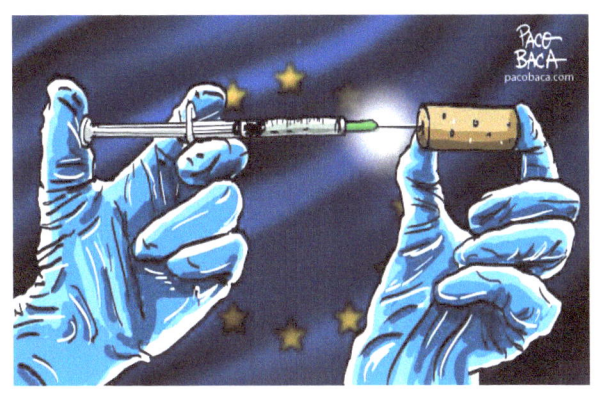

September 8-9, 2020

The Review Committee on the operation of the International Health Regulations (2005) (IHR) during the response to COVID-19 begins its work to evaluate the operation of the IHR during the pandemic and recommend the changes it deems necessary

September 10, 2020

The Director General of the WHO, and Dr. Úrsula von der Leyen, President of the European Commission, jointly host the inaugural meeting of the Facilitation Council of the Accelerator for access to tools against COVID-19 (ACT Accelerator). The meeting is co-chaired by Mr. Cyril Ramaphosa,

President of South Africa, and Ms. Erna Solberg, Prime Minister of Norway.

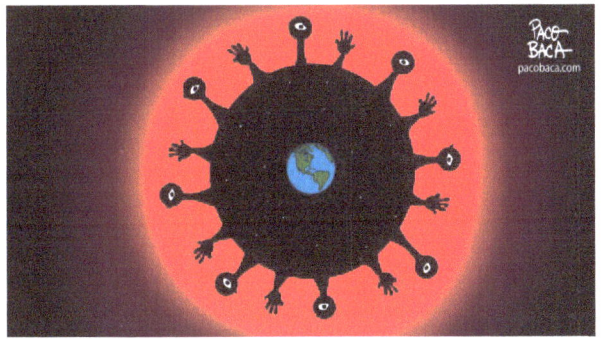

September 11, 2020

WHO publishes interim guidance highlighting the value of antigen-based rapid diagnostic tests for the SARS-CoV-2 virus, in areas where community transmission is widespread and amplification-based diagnostics are not available of nucleic acids or where test results are significantly delayed.

September 14, 2020

The Global Preparedness Monitoring Board, an independent monitoring and accountability body to ensure preparedness for global health crises, publishes its report A World in Disarray.

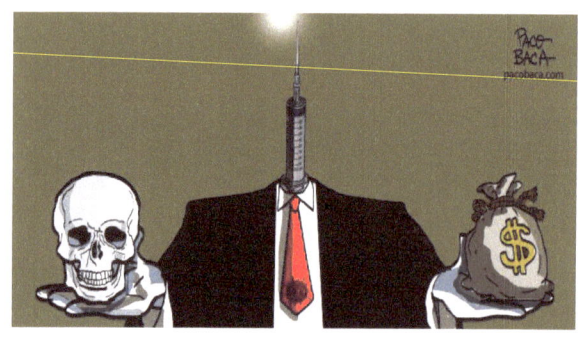

September 15, 2020 – October 2, 2020

At the unprecedented virtual high-level meeting of the seventy-fifth session of the United Nations General Assembly, WHO calls on world leaders to support the Access to COVID-19 Tools Accelerator

(ACT Accelerator), keep up the momentum towards achieving the Sustainable Development Goals and prepare to face the next pandemic together.

September 17, 2020

Independent Pandemic Preparedness and Response Group Holds First Meeting, Faced with Unprecedented Exposure of Healthcare Workers to Risks From COVID-19, WHO Issues Letter on World Safety Day of the Patient calling for measures to protect health workers.

September 21, 2020

The WHO Strategic Advisory Group of Experts on Immunization (SAGE) publishes interim guidance on influenza vaccination during the COVID-19 pandemic.

As of that date, 64 high-income countries and territories have signed up to the COVAX Facility, a global initiative that brings governments and manufacturers together to ensure that future COVID-19 vaccines reach those most in need, By that date , 156 countries and territories, representing approximately 64% of the world's population, have already committed or are eligible to receive vaccines through the COVAX Facility.

September 22, 2020

WHO publishes the first Emergency Use List of a quality antigen-based rapid diagnostic test to detect the SARS-CoV-2 virus, which causes COVID 19.

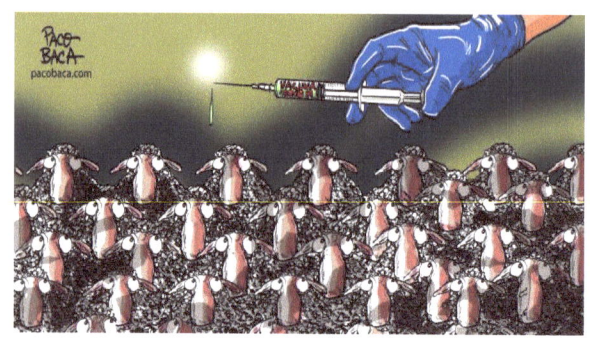

September 23, 2020

WHO issues joint statement with other UN partners and the International Federation of Red Cross and Red Crescent Societies (IFRC) calling for action to manage the COVID-19 'infodemic' 19 from an overabundance of information, both online and offline.

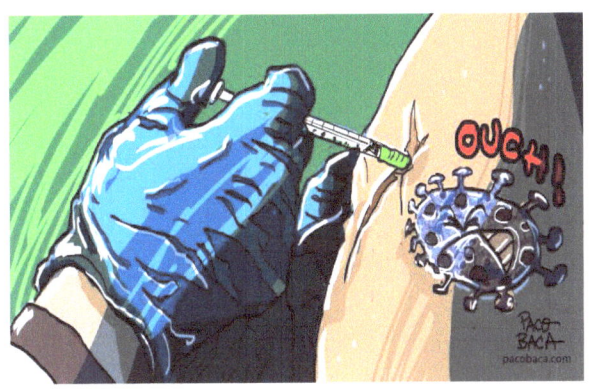

September 24, 2020

The Access to COVID-19 Tools Accelerator (ACT Accelerator) publishes the case for financial investment and funding requirements, covering September 2020 to December 2021.

September 28, 2020

WHO joins with various partners to make 120 million affordable, quality rapid tests for COVID-19 available to low- and middle-income countries.

September 30, 2020

The United Nations and partners welcome approximately $1 billion in new funding for the Access to COVID-19 Tools Accelerator (ACT Accelerator), coming from governments, the private sector, civil society and others international organizations.

October 1, 2020

WHO opens a call for expressions of interest for manufacturers of vaccines against COVID-19 who wish to submit applications for prequalification and/or inclusion in the Emergency Use List.

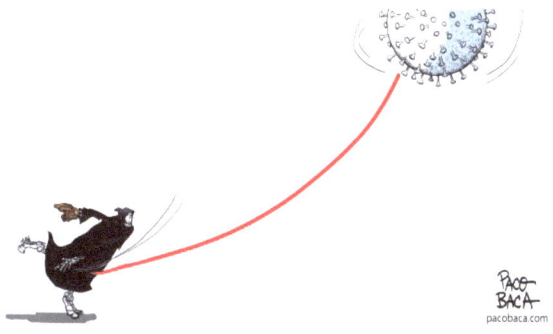

October 5, 2020

WHO reports survey results showing that the COVID 19 pandemic has disrupted or paralyzed essential mental health services in 93% of the 130 countries studied, while demand for mental health services increases.

5 - 6 October 2020

An extraordinary meeting of the Executive Board on the response to COVID 19 is held, presenting an update on the implementation of resolution WHA73.1 adopted by Member States at the World Health Assembly held in May 2020

October 6, 2020

The WHO Foundation, UN75 and Christie's present The Future is Unwritten 'Healing Arts' initiative, a call for cultural action to support a global response to COVID-19 through the arts. The initiative aims to raise awareness of a global path to recovery and raise critical funds to mobilize artists and healthcare professionals in support of the most vulnerable and at-risk communities and health systems. weaker.

October 13, 2020

WHO publishes a joint statement with the International Labor Organization, the Food and Agriculture Organization of the United Nations and the International Fund for Agricultural Development, calling for urgent and ambitious measures to mitigate the effects of COVID-19 on livelihoods, health and food systems.

October 14, 2020

The WHO publishes its World report on tuberculosis for 2020, which highlights a considerable reduction in the notification of tuberculosis cases, but also the efforts of countries to mitigate the impact of the pandemic.

October 15, 2020

WHO announces conclusive evidence on the efficacy of repositioned drugs for COVID-19. Interim results from the Solidarity trial indicate that therapeutic regimens of remdesivir, hydroxychloroquine, lopinavir/ritonavir, and interferon appear to have little or no effect on 28-day mortality or hospital outcomes in COVID-19 patients.

October 19, 2020

WHO partners with musician Kim Sledge and social impact company The World We Want to launch the #WeAreFamily campaign, aimed at inspiring global solidarity for better health. Part of the proceeds from a special edition of the cover of the song "We Are Family" will be donated to the Foundation for WHO.

October 22, 2020

WHO and the Wikimedia Foundation (the non-profit organization that hosts Wikipedia) announce a collaboration to expand open access to the latest and most reliable information on COVID-19.

25 - 27 October 2020

At the World Health Summit, the director-general and several WHO managers and experts call for increased investment in innovation, research and solutions related to COVID-19.

October 29, 2020

The Director General convenes for the fifth time the Emergency Committee of the International Health Regulations on COVID-19. The Committee considers that the pandemic continues to constitute a Public Health Emergency of International Concern (ESPII) and offers its advice to the Director General.

The CEO declares that the COVID-19 outbreak continues to constitute a PHEIC. Accepts the Committee's advice to WHO and forwards it to States Parties as temporary recommendations under the International Health Regulations (2005).

November 4, 2020

The Independent Advisory and Oversight Committee for the WHO Health Emergencies Program releases a report on reforming WHO's work on outbreaks and emergencies in 2016.

November 5, 2020

The WHO publishes the mandate for the global study of the origins of SARS-CoV-2, convened by the WHO.

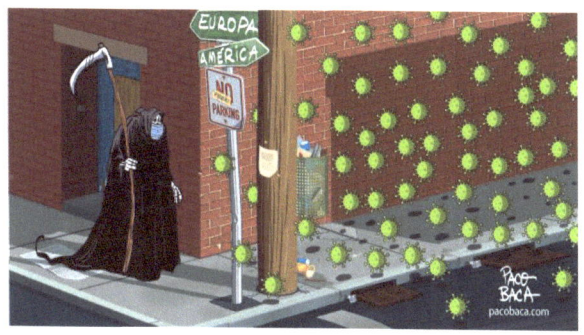

November 6, 2020

WHO publishes in Outbreaks a report on a variant strain of SARS-CoV-2 associated with mink in Denmark. The report contains an overview of the Danish public health response, as well as risk assessment and advice from the WHO.

November 9 - 13, 2020

The resumption of the 73rd World Health Assembly is being held virtually.

The Assembly adopts resolution EB146.R10 aimed at strengthening preparedness for health emergencies.

In recognition of the dedication and sacrifice of the millions of health and healthcare workers on the front lines of the COVID-19 pandemic, Member States unanimously designate 2021 as the International Year of Workers Sanitary and Assistance.

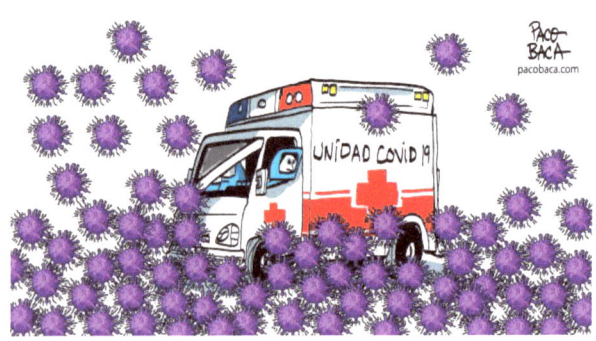

November 10, 2020

WHO launches the "We Are #InThisTogether" campaign to promote collaboration and adherence to five key measures to combat COVID-19: wash hands, wear a mask, cough and sneeze safely, keep your distance and open the windows.

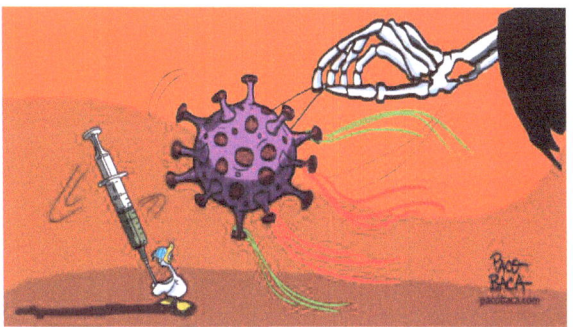

November 11 - 13, 2020

At the Paris Peace Forum, the European Commission, France, Spain, the Republic of Korea and the Bill & Melinda Gates Foundation pledge US$360 million to the COVAX Facility, the immunization pillar of the ACT Accelerator. The CEO undersigns a letter to G20 leaders co-signed with fellow COVID-19 Tool Access Accelerator advocates: Cyril Ramaphosa. President of South Africa, Erna Solberg, Prime Minister of Norway, and Ursula Von der Leyen, President of the European Commission, highlighting the need for an immediate investment of US$4.5 billion for vaccines.

November 18, 2020

The WHO Academy launches its first augmented reality course for health personnel on the proper use of personal protective equipment against COVID-19.

November 19, 2020

WHO provides Member States with updated information on its investigations into the origin of the virus, including the composition of the international team and collaboration with their Chinese counterparts, and publishes the list of the international team.

November 20, 2020

WHO publishes guidance on therapeutic options and COVID-19, with new information for clinicians including a conditional recommendation against the use of remdesivir in hospitalized patients with COVID-19, regardless of disease severity.

November 21, 2020

The Director-General addresses the G20 Leaders' Summit and calls for action to ensure that COVID-19 vaccines are allocated fairly as global public goods, the International Health Regulations are fully implemented, address the vulnerabilities and inequities at the root of the pandemic and help fill the Accelerator funding gap.

December 8, 2020.

The WHO overview of COVID-19 vaccine candidates includes 52 candidate vaccines in clinical evaluation and 162 in preclinical evaluation.

Vaccines, confinement and healthy distance mark the end of the first year in the COVID 19 Pandemic.

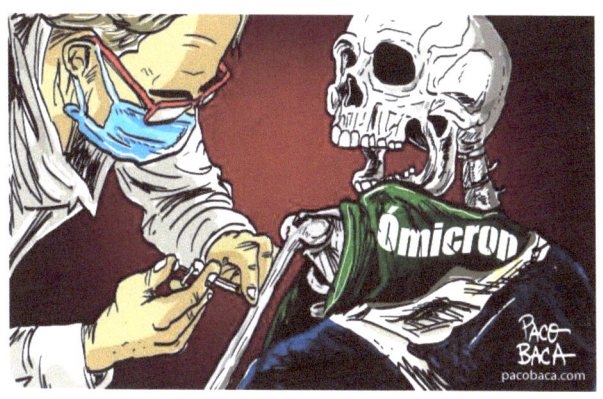

What you need to know.

New variants of SARS-COV-2, the virus that causes COVID-19 will continue to occur.
CDC coordinates collaborative partnerships which continue to fuel the largest viral genomic sequencing effort to date.

The Omicron variant, which emerged in November 2021, has many lineages.

New lineages continue to emerge and spread in the United States and globally.

Conclusion.

This 2020, the world changed and there is already -normality- talk of this "New Normality", which implies that, as of 2021, the implementation of the -welfare state- and the introduction of substantial changes and guidelines will be established. programmed in the world health policy, listed in a protocol for compliance with the objectives of the 2030 agenda. Which, paradoxically, begin with the regulation of mechanisms implemented to control vaccination with the strategies deployed during the 2020 pandemic. All this within of the process of establishing a New World Order of Nations.

"The immunization 2030 Agenda" (AI2030) sets out a global, ambitious and global vision and strategy for vaccines and immunization during the decade 2021-2030. It builds on lessons learned, recognizes the persistent and new challenges posed by infectious diseases, and seizes new opportunities to address them. AI2030 places immunization as a key factor in respecting the fundamental right of people to enjoy the highest attainable standard of physical and mental health and also as an investment in the future through the creation of a healthier, safer and more prosperous world for all. all. It aspires to ensure that we maintain the progress achieved with so much effort and also that we achieve more without leaving anyone behind, in any circumstance or stage of life.

AI2030 is designed to inspire and harmonize the activities of community, national, regional and global stakeholders: national governments, regional bodies, global agencies, development partners, health professionals, research and university institutions, developers and manufacturers. of vaccines, the private sector and civil society. Its impact will be maximized by making more effective and efficient use of resources, innovating to improve performance, and implementing measures to achieve financial and programmatic sustainability. Success will depend on building and strengthening partnerships within and outside the health sector in a coordinated effort to improve access to high-quality, affordable primary health care, achieve universal health coverage, and accelerate progress. towards achieving the 2030 Sustainable Development Goals (SDGs). AI2030 provides a long-term strategic framework to guide a dynamic implementation phase that responds to evolving country needs and the global context over the next decade"

-2030 Immunization Agenda- A global strategy to leave no one behind-
World Health Assembly.

The reality surpasses us day by day.

Without us realizing. We are surprised. It takes us away.

And leaves us paradise within chaos.

Reference sources.

www.who.org

Rambaut,A.Virological.org

Nature

Twitter

Newspapers of the time.

World Health Assembly.

CAS Article Google Scholar.

Group of Experts on Strategic Advice on Immunization. The Global Vaccine Action Plan 2011–2020. Review and lessons learned. Geneva, World Health Organization; 2019 accessed in March 2020.

Group of Experts on Strategic Advice on Immunization. The Global Vaccine Action Plan 2011–2020.

Johns Hopkins University, International Vaccine Access Center. Methodology report: decade of vaccines economics

(DOVE). Return on investment analysis. Medford (MA): Immunization Economics; 2019

Quantitative risk assessment of the effects of climate change on selected causes of death, 2030s and 2050s. Geneva, World Health Organization, 2014.

Thanks to all.

All rights reserved in favor of its author.

Paco Baca Copyright 2021.
2020
The Plandemic year.

www.ingramcontent.com/pod-product-compliance
Lightning Source LLC
Chambersburg PA
CBHW040359220526
45473CB00025B/2407